MIND FOR MATH:
BE A HUMAN CALCULATOR

*Calculate Sums At Lighting Speed,
Think Quickly, Clearly, Focus Fast And
Get The Results You Desire*

(WHILE STUDYING LESS)

© Copyright 2018 by ___Tim Perse_____ - All rights reserved.

This document is geared towards providing exact and reliable information in regards to the topic and issue covered. The publication is sold with the idea that the publisher is not required to render accounting, officially permitted, or otherwise, qualified services. If advice is necessary, legal or professional, a practiced individual in the profession should be ordered.

- From a Declaration of Principles which was accepted and approved equally by a Committee of the American Bar Association and a Committee of Publishers and Associations.

In no way is it legal to reproduce, duplicate, or transmit any part of this document in either electronic means or in printed format. Recording of this publication is strictly prohibited and any storage of this document is not allowed unless with written permission from the publisher. All rights reserved.

The information provided herein is stated to be truthful and consistent, in that any liability, in terms of inattention or otherwise, by any usage or abuse of any policies, processes, or directions contained within is the solitary and utter responsibility of the recipient reader. Under no circumstances will any legal responsibility or blame be held against the publisher for any reparation, damages, or monetary loss due to the information herein, either directly or indirectly.

Respective authors own all copyrights not held by the publisher.

The information herein is offered for informational purposes solely, and is universal as so. The presentation of the information is without contract or any type of guarantee assurance.

The trademarks that are used are without any consent, and the publication of the trademark is without permission or backing by the trademark owner. All trademarks and brands within this book are for clarifying purposes only and are the owned by the owners themselves, not affiliated with this document.

FORWARD

Thank you for purchasing **Mind For Math!**
A book created to make people aware of just how easy multiplication can be.

You may be wondering what type of mathematics this book demonstrates. It is an ancient form of mathematics known as Vedic Mathematics which is over 3000 years old. It has only recently been discovered due to the fact that the manuscripts containing the detailed workings of Vedic Mathematics were destroyed. This e-book does not bring into play the original names of these mathematical rules, which are called Sutras as it aims to keep the information as clear and as straight forward as possible.

The e-book is designed to be read page by page, in order. By the end of the e-book you should be doing calculations in your head easily and with no trouble at all!

Enjoy the e-book and use it as your reference tool to multiplication. Encourage friends to explore Vedic Mathematics and demonstrate to them just how easy it is to do mental multiplication. Then again, don't tell them and let them think you are a human calculator!

CONTENTS

10 times tables 2

20 times tables 9

Numbers near 100 15

Vertically and Crosswise 19

Multiplication shortcuts 23

Multiplying by 11 shortcut 26

Multiplying by 9's 30

5 squared shortcut 38

Squaring two digit numbers 40

Decimal multiplication 44

Tips and Tricks 48

Checking answers 52

Summary so far 54

Dividing 9 into two numbers 55

Dividing 9 into more than two numbers 58

Dividing by 7 62

Dividing by 8 64

Simple Division 65

Adding Fractions 68

Subtracting Fractions 70

Multiplying Fractions 71
Dividing Fractions 72
Square Roots 74
Estimating Square Roots 76
Quick Sums 78
Conclusion 82

MIND FOR MATH! MULTIPLICATION

Numbers play an important role in everybody's life. You can't help but deal with them every day whether you want to or not.

This book aims to break down the mental barrier between you and numbers. Sure you might be good at your tables and don't mind the method you use to find the answer, to say 7 X 8

But what if there is a quicker way? A way that will make sure you get the right answer ever time? A way that can give the answer to multiplying anything by 9999 in a matter of seconds?! And the beauty of it all is that it is easy!

What if in ten minutes you will be able to multiply anything up to 20 X 20 confidently and correctly and that you will be amazed just how easy the whole method is? You will wonder why you were never taught this method in school!

Well, the good news is that you will be able to leave that old method behind and be able to move onto bigger more "difficult" sums in a relatively short period of time.

And the best thing of all is that you will enjoy the process!

This e-book will show you how to learn up to 100 X 100 and then be able to perform larger multiplications using the methods you will learn. Amaze your friends with your newly found mathematical abilities and become the class genius!

On working with this book it is assumed you can calculate basic times tables such as 2 X 3 and 4 X 4. It is from these basic calculations that you can build on making your in built calculator work more effectively.

TEN TIMES TABLES

Learning up to 10 X 10 was a major stumbling block for many people in school. Sitting there reciting the tables hoping that repetition would eventually get the numbers to stick in the brain.
Maybe the numbers did sink in, maybe they didn't. When you learn this technique it won't matter whether they sank in or not!

This whole multiplication method is built around a "base". A base is a number that we can use as a reference point to make our multiplication easier.

Let's try this base out.

For our first multiplication we will try 8 X 7

Using your old multiplication method you would probably recite 8, 16, 24, 32, 40, 48, and 56 and count these numbers on your fingers until you get to the 7th finger.

Using the new method you won't have to count on your fingers any more.

8 X 7

We need a common base between the numbers 8 and 7. What whole number are they close to?

10 is the closest to 7 and 8.

> **10 is called a "base" or reference number. If it is confusing you, call it what you like. Just remember that you need a number that is close to the numbers being multiplied. In this case it is 10.**
>
> **If we were calculating 12 X 13 we can also use ten as it is close to them.**

Ok, back to the sum 8 X 7

We use the base 10 so lay the sum out like this:

$$(10) \quad 8 \times 7$$

How many numbers does it take to make the 8 into 10? Two.
How many numbers does it take to make the 7 into 10? Three

So put these numbers *below* the sum like so:

$$(10) \quad 8 \times 7$$
$$\quad\quad\quad\; \mathbf{2 \quad 3}$$

So why are we putting the numbers *below* the sum? This is because the numbers being calculated are below the base number, ten. As the numbers are below the base they carry a negative value.

So what's next? Now we have to subtract *diagonally* either 7 – 2 **OR** 8 – 3. It doesn't matter which one you choose to subtract, just make sure it is the easiest one. In this example subtracting 2 from 7 seems to be the easier choice.

So, 7 – 2 = 5

5 is the first part of the answer. Multiply this 5 by the base number of 10, giving 50.

So how do we get the second part of the answer?

Let's look back at the sum again:

$$(10) \quad 8 \times 7 = 5?$$
$$2 \quad\; 3$$

We now need to find the second part of the answer.
To do this we multiply the two numbers that make up 10. In this case it is the 2 and 3

2 X 3 = 6

6 is the second part of the answer. Put the two digits together and our sum is complete:

8 X 7 = 56
Let's try another. 6 X 12

$$(10) \quad 8 \times 7 = 56$$
$$2 \quad\; 3$$

Again we need to find a base number. 10 is close to both digits.

$$(10) \quad 6 \times 12$$

Now find the number of digits it takes to make up ten.

$$2$$
$$(10) \quad 6 \times 12$$
$$4$$

So why is the 2 above the 12? The reason for this is because 12 is 2 above ten. So we put it above the 12. Six is 4 below ten so we put the four below the 6.

Next step we add or subtract diagonally. The reason why you may add is because the 2 is above ten giving it a positive value.

So you could add or subtract like so:

$$6 + 2 = 8 \text{ OR } 12 - 4 = 8$$

So we have 8. We multiply this 8 by the base number 10, giving 80.

$$\begin{array}{c} 2 \\ (10) \quad 6 \text{ X } 12 = 8? \\ 4 \end{array}$$

So how do we get the next number, seeing as one number is below and the other is above?

Simple. Multiply the 4 by 2. Note here that as the four is below, it has a – (minus) value.

So the sum is -4 X 2 = -8

What about before when the two numbers were below? Why were they not negative? The reason for this is because:

> **A minus times a minus = a plus**
> **A minus times a plus = a minus**
> **A plus times a plus = a plus**

A good way to remember that minus times a minus is a plus is imagine a minus sign is a sword. Picture in your mind two swords clashing together. They look like a cross, or a plus sign on its side.

For minus times a plus is a minus imagine again the sword knocking the cross out of the way leaving just the one sword, a minus.

Ok so now we have to finish off the sum.

$$\begin{array}{c} 2 \\ (10)\ \ 6 \times 12 = 8? \\ 4 \end{array}$$

Remember we have 80 as we multiplied the 8 by the base number. So it is now 80 – 8 = 72.

Therefore our completed sum becomes 6 X 12 = 72

Instead of subtracting 8 from 80 you can subtract 10 and add 2.
80 – 10 = 70 + 2 = 72

Let's go one more.

14 X 17

Again we can use ten as our base number.

$$\begin{array}{c} 4\ \ \ \ 7 \\ (10)\ \ 14 \times 17 = \end{array}$$

So again what we do is find the difference between 14 and 10 and the difference between 17 and ten. The numbers being calculated are above the base so the difference goes on top of the sum.

Next stage is to **add** diagonally. Why add? Well both numbers are above 10 so they both have a positive value.
So either 14 + 7 **OR** 17 + 4
Here you can add ten to 14 giving 24 and then subtract 3 giving 21. This way may be easier for you.

So now we have 21. Remember to multiply this 21 by the base number 10 giving 210.

$$\begin{array}{c} 4\ \ \ \ 7 \\ (10)\ \ 14 \times 17 = 21? \end{array}$$

How do we find the next number? Multiply the difference, 4 and 7 by each other.

4 X 7 = 28

But how do we do 4 X 7? Use the exact same method you are using to multiply 14 X 17!

So now we have 210 and 28
Add these two numbers together to get our final answer.
To simplify this sum, just add 30 to 210 giving 240 and then subtract 2 giving the answer 238.

14 X 17 = 238

You may be wondering about numbers in and around 5. Can we use a base for these? Using a base of ten will just make our sum require the same amount of work as the sum we want to work out, so we could use a base of five.

Let's try 4 X 6
Base to use here is 5.

$$(5) \quad 4 \times 3$$
$$ 1 \quad 2$$

Subtract 1 from 3 or two from 4 giving 2
The easiest way to use 5 as a base is to multiply by ten and then divide by two. So 2 X 10 = 20 and then divide by 2 giving ten. Multiply the differences from the base, 1 and 2 together giving two and add this to ten. 4 X 3 = 12.

If we were to use 10 as our base our work would look like so

$$(10) \quad 4 \times 3$$
$$ 6 \quad 7$$

So as you can see we would be multiplying 6 X 7 together making out sum longer! Using a base of five lets us take the differences from 5 instead of 10.

Ten times ten questions

a) 8 X 7   d) 7 X 7   g) 9 X 16

b) 11 X 8   e) 13 X 14   h) 18 X 13

c) 12 X 4   f) 15 X 6

Answers:
a) 56
b) 88
c) 48
d) 49
e) 182
f) 90
g) 151
h) 234

TWENTY TIMES TABLES

Being able to calculate 17 X 24 on paper may prove difficult enough besides being able to calculate it in your head! Hopefully by now the method you have learnt for 10 X 10 has made you inquisitive and left you wanting to know how to calculate bigger numbers.

Well you will be happy to hear that 20 X 20 is not difficult at all. Let's take a look at 17 X 24

First, like 10 X 10 we need a reference number. Both numbers are close to 20 so twenty will make an ideal reference number.
Set out the sum like we did for 10 X 10

$$(20) \quad 17 \text{ X } 24$$

Again, how many numbers is the difference between 17 and 20, and also 24 and 20? Twenty – 17 = 3 and 24 – 20 = 4
Place these numbers as follows:

$$\begin{array}{c} 4 \\ (20) \quad 17 \text{ X } 24 \\ 3 \end{array}$$

Notice that 3 is below and 4 is on top, as 17 is less than twenty by 3 and 4 is more than twenty by 4.

Like before you can either do 17 + 4 = 21 **OR** 24 – 3 = 21

Now before when working with 10 X 10 we would multiply this number by 10, our reference number. But our reference number

has changed. It is now 20. So we need to multiply 21 by twenty. To do this just double 21 and add a 0 on the end.
21 doubled is 42 and add a 0 becomes 420.

$$\begin{array}{c} 4 \\ (20) \quad 17 \times 24 = 42? \\ 3 \end{array}$$

Our last step is to calculate the last part, -3 X 4 = -12

So instead of adding twelve to 420 we subtract.

An easy way to do this is taking ten from 420 and then two from it.
420 – 10 = 410 – 2 = 408.
Therefore 17 X 24 = 408

Let's try 23 X 26

Again lay out the sum like so:

$$(20) \quad 23 \times 26 = ?$$

Now find the difference between 23 and 20 and also 26 and twenty. 23 – 20 = 3 and 26 – 20 = 6

$$\begin{array}{cc} 3 & 6 \\ (20) \quad 23 \times 26 = ? \end{array}$$

Ok, just like before in all the examples we add or subtract. Both numbers are above 20 so we can add diagonally either 23 + 6 = 29 **OR** 26 + 3 = 29

Then as we are using 20 as a reference number we have to multiply 29 by 20.

To do this, round 29 to 30 and double it. 60. Take two away. 60 – 2 = 58 and then add 0: 580

So now it looks like this:

$$\begin{matrix} & \mathbf{3} & \mathbf{6} \\ (20) & 23 \times 26 & =58? \end{matrix}$$

Our final step is to find the last part of the sum. Just like before we multiply the differences together. $6 \times 3 = 18$.
Add 18 to 580. To do this add 20 and take 2 away.
$580 + 20 = 600$
$600 - 2 = 598$

$23 \times 26 = 598$

So what about multiplying 8×28?

We could use 20 as the base number here but the multiplication gets more difficult.
For example we would need to multiply 12×8 in this sum, but this is not making our sum any easier is it?

So how about if we use *two* base numbers? Don't worry it's not difficult to use two base numbers. It's just one more step in the sum.

Choosing which base numbers to use is not difficult to do. You do this by looking at the numbers in the sum and deciding which bases the numbers are nearer to.

8×27

We can use 10 and 20 as our bases.

But notice 27 is nearer to 30 so let's use thirty as our second base. Our sum now looks like this:

$$\begin{matrix} (10 \times 3) & 8 \times 27 & =? \\ & \mathbf{2} \quad \mathbf{3} & \end{matrix}$$

So what is (10 X 3) all about? Well we need to include a base of 30 and

10 X 3 = 30 so we have our base of 10 and our base of 30 included in the sum.

$$(10 \text{ X } 3) \quad 8 \text{ X } 27 =?$$
$$\mathbf{2} \quad \mathbf{3}$$

So our next step is to work out the difference between our sum and the base. What is the difference between 10 and 8? 2. 30 and 27? 3. Both numbers are below their references so we put them below.

Now the only difference here is that we take the difference of the number that uses 10 as a reference and we multiply it by 3, our other reference.
So 2 X 3 = 6
Now lay out the sum as follows:

$$(10 \text{ X } 3) \quad 8 \text{ X } 27 =?$$
$$\mathbf{2} \quad \mathbf{3}$$
$$\mathbf{6}$$

Ok as the six was multiplied by 2 which carries a negative value, it too becomes negative, i.e. – 6.
So take 6 from 27
27 – 6 = 21
We then multiply 21 by our reference 10 giving 210 Then we multiply 2 X 3 our differences giving 6 Add this 6 to 210
210 + 6 = 216
8 X 27 = 216

Ok let's try it with a larger number, 21 X 84
So again let's find a base number for each part of the sum. 20 X 4 looks good as 20 X 4 = 80

$$
\begin{array}{cc}
 & 4 \\
1 & 4 \\
\end{array}
$$
(20 X 4) 21 X 84

Like before, we multiply the 1 by 4 giving 4 and place that above the 1. Now the usual procedure, add diagonally.
84 + 4 = 88
Multiply 88 by the base 20. To do this, just double 88 and add a 0. You could double 90 or 100 and subtract the difference. Use 100 as it is easier:
100 doubled = 200
200 subtract 24 (remember, 88 is twelve away from 100 and we are doubling it so it's 24).
To make this easier still just subtract 20 and then 4. 200 − 24 = 176.
Then add a zero, 1760
Now multiply the differences 1 and 4 to give 4 1 X 4 = 4 and then add this to 1760
1760 + 4 = 1764
21 X 84 = 1764

20 times 20 questions

a) 17 X 22

b) 11 X 25

c) 9 X 21

d) 7 X 26

e) 23 X 14

f) 27 X 8

g) 19 X 27

h) 28 X 23

Answers:
a) 374
b) 275
c) 189
d) 182
e) 322
f) 216
g) 513
h) 644

MULTIPLYING NEAR 100

So can what we learnt for multiplying numbers below 100 work for numbers above 100? Of course it can! And the good news is that it is easy too!

Let's try a sum:

102 X 107

Ok now before reading this e-book you would never have thought that doing this type of sum was easy, well hopefully by now you are beginning to see just how easy they can be!
Again we need a reference number. What looks like it would suit here? 100 does. So that's our reference number.

$$(100) \quad 102 \text{ X } 107$$

Ok, just like multiplying numbers below 100 we have to find the difference between the base and the numbers in the sum.

Lay out the sum again like before:

$$\begin{array}{cc} 2 & 7 \\ (100) \quad 102 \text{ X } 107 \end{array}$$

Now as we are only working with one base we don't need to multiply the 2 above 102 with another reference.
Add diagonally either 102 + 2 **OR** 107 + 2
So we get 109 and we now need to multiply this by our reference base of 100. To do this just add two zeros to 109: 10900

Finally just multiply the 2 by 7: 2 X 7 = 14 and this fourteen to 10900
10900 + 14 = 10914

There is very little difference in multiplying large numbers and lower numbers just what reference number to use.

So how about 96 X 113

We can use 100 as our base number here.

$$(100) \quad 96 \text{ X } 113$$

Find the difference between the numbers in the sum and the reference number:

$$\begin{array}{c} 13 \\ (100) \quad 96 \text{ X } 113 \\ 4 \end{array}$$

Remember 96 is four less than 100 so it has a negative value!

In this case we can either add or subtract diagonally, either 96 + 13 **OR** 113 − 4. Either way they equal the same number. 113 − 4 = 109
Multiply 109 by our base 100 (remember just add two zeros. If we were multiplying by 200 we would double it and add two zeros).
10900
Then multiply the difference of our two sums, 4 and 13. 13 X -4 = -52
Note that the 4 was below so it carries a negative value. So now we add the two values together:
10900 + -52
Take 100 away and add 48.
96 X 113 = 10848

*Here is a good way to subtract 52 from 100.

> **Take all from nine and the last from ten.**

What this means is take for example our 52. Take all from nine, meaning take the 5 from nine giving us 4 and then the last from ten. In this case the last is 2. So 2 from ten is 8. This method works when subtracting from any whole number.

1000 – 546

All from 9 and the last from 10

$$\begin{array}{ccc} 5 & 4 & 6 \\ \downarrow & \downarrow & \downarrow \\ 4 & 5 & 4 \end{array}$$

So 1000 – 546 = 454

So how about 114 X 723?

Again we need to find a reference. 100 is good for 114 but not for 723. So we can use (100 X 7) as it equals 700.

(100 X 7)   114 X 723

Again find the differences between the numbers and their bases;

$$\begin{array}{cc} 14 & 23 \end{array}$$
(100 X 7)   114 X 723

Just like we have done before, using the method we have learnt already multiply 14 by 7, the reference for 723.

14 X 7 = 98

Add 98 to 723

723 + 98 = 821

Multiply 821 by our base 100, in this case just add two zeros. If our base was 300 we would times it by three and add two zeros.

821 X 100 = 82100

Next multiply 14 by 23.

14 X 23 = 322

Finally we add 322 to 82100

82100 + 322 = 82422

We can see in this example that there was a little more work involved in this sum. But that does not mean it was any more difficult. In the above sum we needed to multiply 14 X 23. We would have done that in exactly the same way as we would have done for multiplying numbers below 100. That is, we would have used 10 X 2 as our base and reference. Remember, it is not any more difficult, it just may take a little longer to do.

VERTICALLY AND CROSSWISE

The vertically and crosswise method demonstrates another way to multiply numbers together. It is best used when you are unable to determine what bases to use when multiplying.

Let's take a look at how this works

23 X 87

Lay the sum out like so

$$\begin{array}{c} 23 \\ 87 \end{array}$$

The first thing we need to do is multiply vertically 3 and 7

$$\begin{array}{cc} 2 & 3 \\ & | \\ 8 & 7 \end{array}$$

So 3 X 7 = 21

$$_21$$

Two is carried number

Carry over and add to the next

Now we need to do the crosswise part, which is 7 X 2 *plus* 8 X 3

$$_40\ _21$$

And finally we do the vertically part again, to 8 and 2 8 X 2 = 16

```
        2    3
        |
        8    7
       ─────────
So 2 X 8 = 16
       20₄ 0₂ 1
```

23 X 87 = 2001

This method also stands true when multiplying by three numbers. Let's take a look

466 X 734

Lay the sum out like so

$$466$$
$$734$$

The first thing we need to do is multiply vertically 6 and 4

```
            4   6   6
                    |
            7   3   4
           ─────────
So 6 X 4 = 24
                   ₂4
```

Now we need to go the crosswise part, which is 6 X 4 *plus* 6 X 3

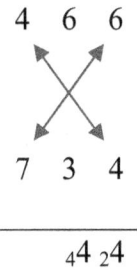

$_44\ _24$

The next step is to continue on with the crosswise part and extend it to all the numbers, like so

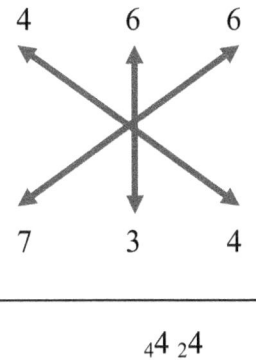

$_44\ _24$

The sum now becomes
4 X 4 plus 6 X 3 plus 6 X 7
16 plus 18 plus 42 = 76 and then add the carried over four from the sum above giving 80

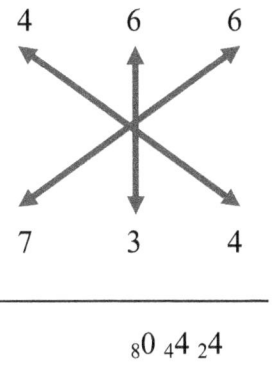

$_80\ _44\ _24$

Again we continue on with the crosswise, this time we use 4 X 3 *plus* 6 X 7

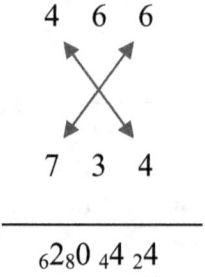

$6_28 0\ _44\ _24$

Finally our last step is to do the vertically part, 4 X 7

```
        4   6   6
                |
        7   3   4
```

So 4 X 7 = 28

34 $_6$2 $_8$0 $_4$4 $_2$4

466 X 734 = 342044

MULTIPLICATION SHORTCUTS

By now you should be finding these sums easy. That's because the method you have just learnt makes them easy. Well you are about to learn another way of making certain multiplications even easier!

This next method works when the numbers being multiplied have the same difference between then from the base reference number.
For example take 17 X 23
The base to be used here would be 20.

There is a difference of 3 between 17 and twenty and also another difference of three between 23 and 20.
So for this next method to work, the differences must be the same.

So how do we do the sum when using this method?
Simple. We take the base; in this case it's 20. We square it, 400 and then subtract the square of the difference between the numbers; in this case 3 squared which is 9.
400 – 9 = 391

But how do we square 20? Just square the 2 which is 2 X 2 and add two zeros on the end!

How about 146 X 154?
Well they both have a difference of four from the base of 150 so we would square 150 and subtract the square of four.
To square 150 just square 15 which is 15 X 15 and add two zeros on the end, giving 22500
Next you subtract the difference of 4 squared, which is 16. 22500 – 16, to do this take away 20 and add four giving 22484.

Multiplying numbers that add to ten

This method allows the person to multiply any two numbers that have the same base and the last digits add to ten.

For example let's look at 24 X 26

You can see that both numbers have the same base, in this case it is 20. The last two numbers add up to ten, 4 plus 6 = 10.

Well then you will be happy to know that there is an even quicker method of multiplying sums like these and you don't need to use the base number for it. This is how it works

24 X 26

We take the first number 2 and multiply it by one more than two. One more than 2 is three so it's 2 X 3 = 6. Finally we multiply the 4 (from the 24), and the 6 (from the 26), together giving 24. Bring the numbers together

24 X 26 = 624

Let's try 41 X 49

Like before, we multiply the base 4 by one more than it, 5 giving 4 X 5 = 20 Then we multiply the 1 (from the 41), and 9 (from the 49) together giving 9. Bring the numbers together

41 X 49 = 2009

Remember though that this 9 is in fact 09 as we need two digits after the 20. Think about it, when using this method whether the two numbers add up to ten, the only number when multiplied together that is less than ten is 1 X 9

If we were to do 2 X 8 (remember they must add to ten and have the same base) we get sixteen. So if the sum was 42 X 48 the answer would be 2016!

42 X 48
Multiply the 4 by one more than it, 5 giving 4 X 5 = 20
2 X 8 = 16
Bring the numbers
together 42 X 48 = 2016

You shouldn't be surprised to learn that this method will also work with three digits

108 X 102
Take the 10 in 108 and multiply it by one more than it 11 giving 10 X 11 = 110
Multiply the 8 and 2
8 X 2 = 16
Bring the numbers
Together 108 X 102 =
11016

Try out a few out for yourself and be amazed at how simple they are!

MULTIPLYING BY 11 SHORTCUT

This neat little trick will show you how to multiply any two numbers by 11 in seconds!

Let's take, for example, 63 X 11

You take the number being multiplied and you split the numbers apart, like so

$$6 \longleftrightarrow 3$$

The next step is to add the two numbers together and put that answer in between the two numbers,
6 + 3 = 9

$$6\ 9\ 3$$

So 63 X 11 = 693
How easy was that?!

There is a slight variation when doing this trick when the numbers you are adding together are over 10.

Lets try 29 X 11

Again we separate the two numbers like so

$$2 \longleftrightarrow 9$$

Next we add the two together 2 + 9 = 11

But we do not put 11 in between the 2 and 9 giving **2119** as this is **wrong**. We carry over the tens unit of 11 to the 2 and add them together giving 319 So 29 X 11 = 319

So how about numbers that have more than two digits in them, like multiplying 234 X 11?

Well, this can be done just as easily you will be happy to know!

This is how it is done.
Separate out the first and the last digits of the number you are multiplying by 11

$$2 \longleftrightarrow 4$$

We don't forget about the middle number, 3 as this is the number we add to both separated digits:

$$2\,(2+3) \longleftrightarrow (3+4)\,4$$

So what we do here is put down are initial separated number, in this case it is 2, then the sum of 2 + 3 which is 5, then the sum of 3 + 4 which is 7 and finally the last separated number which is four.

234 X 11 = 2574

Let's try 566 X 11

Separate out the first and last, just like in the last example:

$$5 \longleftrightarrow 6$$

Now we add the middle digit 5 to the first and the last.

$$5\,(5+6) \longleftrightarrow (6+6)\,6$$

Both of these numbers go over ten so just be careful that you add an extra number where it is required.

$$5\,(5+6 \text{ and } 6+6)\,6$$

giving

5 (11, 12) 6

here we add the 1 in the 12 to 11, giving twelve.

5 (12, 2) 6

and then we add the 1 of the 12 to the 5 giving

6 (2, 2) 6

then just put all the numbers together 566 X 11 = 6226

Let's do one more example. 193 X 11
Separate the first and last digits

1 ⟷ 3

add nine to the first and to the last

1 (1 + 9) ⟷ (9 + 3) 3

add the sums together

1 (10, 12) 3

move the 1 in the twelve and add it to the 10.

1 (11, 2) 3

take the first 1 in the 11 and add it to the first 1 outside the brackets

2 (1, 2) 3

Finally put all the numbers together 193 X 11 = 2123

Multiplying by 11 questions

a) 12 X 11

b) 45 X 11

c) 342 X 11

d) 546 X 11

e) 837 X 11

f) 32 X 11

g) 88 X 11

h) 445 X 11

Answers:

a) 132

b) 495

c) 3762

d) 6006

e) 9207

f) 352

g) 968

h) 4895

MULTIPLYING BY 9's

9 times tables are looked upon as being difficult by most people and with their current method for calculating the answer it doesn't seem surprising. You are about to learn a technique to simple you will be surprised you didn't know it already!

Multiplying a single number by 9
Let's look at 6 X 9
The way to find the answer for this sum is that same as multiplying any single digit by nine.
Bring the 6 down one, to make 5. How many is five from 9? 4. Put the two numbers together, 6 X 9 = 54

How about 7 X 9?
Again bring the seven down one, making 6. How many is 6 from nine? 3. Bring the two numbers together,
7 X 9 = 63

Multiplying two digit numbers by 9

This method is slightly different but definitely no more complicated. Let's try 24 X 9
We look at the first digit, 2. We then add a one to it, giving 3. We then subtract 3 from 24 giving 21
Now we look at the 4 of 24 and find the difference between that and ten. 6 is the difference.
Bring the 21 and the 6
together. 24 X 9 = 216

Another, 43 X 9
Bring up the 4 to 5, and subtract 5 from 43 giving 38. Look back at the 43 in the sum, what is the difference between the 3 and ten? 7. Bring the 38 and 7 together
43 X 9 = 387

73 X 9?
Bring the 7 up one, to 8. 73 – 8 = 65
Look back at 73, what's the difference between the 3 and ten? 7. Bring the 65 and 7 together
73 X 9 = 657

Multiplying by 9 questions

a) 42 X 9

b) 5 X 9

c) 9 X 9

d) 66 X 9

e) 46 X 9

f) 94 X 9

g) 33 X 9

h) 57 X 9

Answers:

a) 378
b) 45
c) 81
d) 594
e) 414
f) 846
g) 297
h) 513

Multiplying a single digit by 99

Multiplying by 99 is not a lot different from multiplying by nine. Let's start by looking at an example.

4 X 99

Bring the 4 down one, to 3.
As there is only one digit we are multiplying we keep the first 9 as it is. What is the difference between four and ten? 6
Put all the numbers together

$$4 \times 99 = 396$$

Let's take a look at 8 X 99

Bring the 8 down one, to 7.
What is the difference between the eight and 10? Two. We only have one digit to multiply 99 by so we keep the 9.
Bring all the numbers together 8 X 99 = 792

Multiplying two digit numbers by 99

So how do we go about multiplying two numbers by 99?

23 X 99

We bring the 3 of 23 down by one, to 22.
Remember the rule "*all from 9 and the last from ten*"? Well that rule applies here.
We have deducted the one to give us 22. Apply the all from nine rule to 23, giving 7 and 7.
Bring all the numbers
together 23 X 99 = 2277

42 X 99

Subtract one from 42 giving 41. Now apply the all from nine… rule, 42 all from 9 is 5 and 8
Bring all the numbers
together 42 X 99 = 4158

How do we calculate sums with more than two nines? Read on.

Multiplying a single digit by 999

You won't be surprised to learn that multiplying by more than two 9's is as easy as 99.
This is how it's done
4 X 999
Again, as we did for single numbers, bring the number down by one, giving 3.
As we only have one digit we don't touch the first two 9's, 99.
For the last part we find the difference between 4 and 10.
Bring all the numbers
together 4 X 999 = 3996

7 X 999
Subtract one from 7 giving 6.
Find the difference between 7 and 10 giving 3.
Bring them all together,
7 X 999 = 6993

Multiplying two digit numbers by 999

This method is very similar to multiplying by 99.

62 X 999

Subtract one from 62 giving 61.
We have two digits in our sum, 6 and 2. So we leave the first 9 alone and apply the all from 9 and last from 10 rule.
62 gives 3 and 8.
Bring all the numbers together 62 X 999 = 61938

48 X 999

Subtract 1 giving 47. Keep the first nine and apply the all from nine and last from ten rule to 48 giving 5 and 2.
Bring all the numbers together 48 X 999 = 47952

This same method can be applied to three digit numbers

467 X 999

Subtract 1 giving 466
As there are three digits there are no nines left over in the 999. All from 9 and last from ten rule is applied to 467 giving us 533
Bring all the numbers together

467 X 999 = 466533

How about

467 X 99999?

Subtract one from 467 giving us 466
There are three digits and 5 nines so we leave the first two nines alone. Apply the all from 9 and last from ten rule again to 467 giving 533.
Bring all the numbers together
467 X 99999 = 46699533

You can see how easy these are can't you? With a little bit of practice you can do these in your head and in no time at all!

5 SQUARED SHORTCUT

This method is brilliant for squaring any two digit number that ends in five.

Here goes

35^2

From the methods you have learnt before you could tackle this sum like so
(30) 35 X 35
But there is an even easier way than using a base of 30.
Take the 3 and 5^2 and separate them

$$3 \longleftrightarrow 5^2$$

Add one to 3 and you get four. 4 X 3 = 12
Bring the 12 and 5^2 together (5^2 is 25).
12 and 25
Therefore
35^2 = 1225

75^2
Add one to 7 giving 8.
8 X 7 = 56
Bring 56 and 5^2 together
75^2 = 5625

5 squared shortcut questions

a) 45^2	c) 65^2	e) 95^2

b) 55^2	d) 35^2

Answers:

a) 2025
b) 3025
c) 4225
d) 1225
e) 9025

SQUARING TWO DIGIT NUMBERS

By now you shouldn't be surprised to learn that there is a shortcut for squaring two digit numbers.

Let's look at 26^2

Below are three steps you must take to calculate the answer

> **Square the first**
> **2 times the first and last**
> **Square the last**

Ok so let's work through each step
26^2

Square the first, so $2^2 = 4$
2 X first and last, (2 X 2) X 6 = 24
Square the last, $6^2 = 36$

So let's put all the numbers together
4, 24, 36

The last digit in the answer is 6, so add the 3 to the 4 in 24 giving 27. Add the two of 27 to the four giving 6.

You could lay the numbers out one above the other

```
    4
   24
    36
    _____  Add the numbers
  676
```

Another, 57^2

Square the first, $5^2 = 25$
$(2 \times 5) \times 7 = 70$
$7^2 = 49$

Lay the numbers out one above the other

   25
    70
     49
  _____ Add the numbers
 3249

Try some out yourself, it won't take you long to get the hang of them.

Another way to square two digit numbers

You may have enjoyed using the previous method for calculating the square of a two digit number, well there is another method, just as easy, if not easier.

Let's take the same example again, 26^2

The method here is to **add** to the sum whatever number it is above its base. So in this case we add 6 to 26

6 + 26 = 32

The base we were using in this example was 20, so we need to multiply 32 by 20. How? Times 32 by two and add a zero.

32 X 2 = 64 and add a zero, 640

Last part we need to add to 640 the square of the number above the base, 20. In this case 6 is above our base 20 so we square 6 and add it to 640

6^2 = 36

640 + 36 = 676

26^2 = 676

Let's try 42^2

Our base here is 40

Add two to 42 as it is 2 above the base giving 44

Now multiply 44 by 4 giving 176 and add a zero as we multiplied by 40 giving 1760

Now square the number we added to 42 which was 2. Two squared is 4

Bring the numbers together

$42^2 = 1764$

That way was easy wasn't it? But did you notice that 42^2 was near the base 50? So we could have actually used 50 as our base. How? Like so

42^2 and our base is 50 so 42 is eight less than fifty so we **subtract** 8 from 42 giving 34. So how do we multiply 34 by our base 50? Just multiply it by 100 and divide it by 2

34 X 100 = 3400 and then divide by 2 diving 1700

What number did we subtract from 42? We subtracted 8 so we now need to square this number and add it to 1700

$8^2 = 64$

64 + 1700 = 1764

$42^2 = 1764$

DECIMAL MULTIPLICATION

Multiplying a number that has decimal points in it may seem quite difficult to do. But with the methods you have already learnt you will see how easy they are to do.

Let's take 1.4 X 23

So how would we approach this sum? First thing when multiplying with a decimal point in the sum is to just forget about the decimal all together. In this case, 1.4 would become 14. So what we did was just remove the decimal. 3.6 would become 36 and so on.

So can you see what we are going to do now? We look at the sum again and it now looks like this:
14 X 23
So like before we choose a base that is common to both. Ten and two.

$$(10 \times 2) \quad 14 \times 23$$

Again get the numbers that differ from the base.

$$\begin{array}{cc} \mathbf{4} & \mathbf{3} \\ (10 \times 2) \quad 14 \times 23 \end{array}$$

Just like previous examples we multiply the four by the base 2, giving 8. We put this eight above the four.

$$\begin{array}{cc} \mathbf{8} & \\ \mathbf{4} & \mathbf{3} \\ (10 \times 2) \quad 14 \times 23 \end{array}$$

Add the 8 to 23 giving 31. Multiply 31 by the base 10 giving 310.

Just like before multiply the 4 by 3 giving twelve and add this to 310.

12 + 310 = 322.

Ok so this would otherwise be our answer for 14 X 23 but remember the original sum was 1.4 X 23 so how do we figure out what to do with the decimal?

Easy. How many numbers are to the right of the decimal point? The answer is 1. So from the answer 322 place a decimal point one place from the right.

1.4 X 23 = 32.2

If we were multiplying 1.44 X 23 we would place the decimal point two places from the right as there are two numbers to the right of the decimal point in the original sum.

Let's try 2.4 X 3.6

So like the last example we just ignore the decimals completely, giving us 24 X 36 and calculate it like you have learnt.

Use the base 10 and three.

$$(10 \text{ X } 3)\ 24 \text{ X } 36$$

Find the difference from the bases:

$$14 \quad 6$$
$$(10 \text{ X } 3)\ 24 \text{ X } 36$$

Multiply the 14 by the base 3 giving 42

$$42$$
$$14 \quad 6$$
$$(10 \text{ X } 3)\ 24 \text{ X } 36$$

Add this 42 to 36 giving 78 and multiply by 10 our base.
78 X 10 = 780

Multiply 14 X 6 giving 84 and add this to 780 giving 864
24 X 36 = 864

Remember we were originally multiplying two numbers with decimals in them; 2.4 and 3.6

In total, how many digits are there to the right of each decimal point?

With 2.4 there is one and with 3.6 there is one also. Add these two together so there is a total of two decimal points between them.

So looking back at our answer of 864, place a decimal point two digits in from the right, giving us 8.64

So 2.4 X 3.6 = 8.64

Let's try one more example
1.08 X 3.87

So just like before ignore the decimal points for the time being so our sum now looks like this:

108 X 387

Find our bases

100 and three as 100 X 4 will give us 400 which will work out well for 387

$$(100 \text{ X } 4) \quad 108 \text{ X } 387$$

Find the difference between the numbers and their bases

$$\begin{matrix} & 8 & \\ (100 \text{ X } 4) & 108 \text{ X } 387 \\ & & 13 \end{matrix}$$

Now we multiply the 8 by the base 4 giving us 32

$$
\begin{array}{r}
32 \\
8 \\
(100 \times 4) \quad 108 \times 387 \\
13
\end{array}
$$

So now we add 32 to 387 giving us 419
Multiply 419 by the base 100 giving us 41900
Now we multiply 8 by 13 giving us 104 and *subtract* this from 41900
108 X 387 = 41796
Remember we subtract as the 13 we multiplied by 8 carries a negative value as it was 13 below the base 400.
So now we go back to the original sum which was 1.08 X 3.87
How many numbers are to the right of 1.08? Two. How many to the right of 3.87? Two also. Add these together giving 4. So now we know that we have to have four numbers to the right of the decimal point in our answer:
1.08 X 3.87 = 4.1796

What about 10.8 X 3.87? Well there is one number to the right of 10.8 and there are two to the right of 3.87. Add these together giving us three. So we need to have three numbers to the right of the decimal point in our answer 41796
10.8 X 3.87 = 41.796

TIPS AND TRICKS

So what can you do with your new found mathematical abilities? Hopefully you will have found the shortcuts useful and have been applying them to your multiplications.
This section will demonstrate a few tricks that may be useful to you in your every day life.

How to convert Celsius to Fahrenheit
The general formula used to do this sum is far too complicated for what it needs to be. This little trick is far easier to do.

Let's try 11°C. Convert it to Fahrenheit.

First step is to double the Celsius.
Second step is to add thirty.
11 X 2 = 22
Add 30 = 52
11°C = 52° F

What about 22°C?

Double it and add thirty.
22 X 2 = 44
Add 30 = 74
22° C = 74°F

So can we work out Fahrenheit to Celsius?
Just do the reverse!
Let's try 78°F
Subtract 30.
Divide it by two.

78 − 30 = 48
48 halved = 24
78°F = 24°C

That's a nice easy way to figure out the temperature conversion, isn't it? Bear in mind this method does not give an exact temperature conversion but it provides us with an adequate result for practical purposes.

Converting kilograms to pounds

This little tip demonstrates an easy method of converting kilograms to pounds. The multiplying by 11 shortcut is needed to make this conversion simple.

Let's try and convert 80 Kg to pounds.

> Use this method if the Kg ends in a 0 or 5

Double it, giving 160
Divide this by ten. To do this just remove the last zero, giving 16.
Multiply 16 by 11. Use the shortcut for multiplying by eleven; add the 1 and 6 and put the answer in between.
16 X 11 = 176
80 Kg = 176 lbs

Numbers that ended in a five or zero worked well as we were dividing by ten.
The proper conversion method is 1 Kilogram = 2.2 pounds.
So to do this sum we multiply the Kilograms by 0.2 and then by 11.

Let's try 12 Kg.
12 X 0.2 = 2.4
Multiply 2.4 by 11. Add the two and four together and put this number to the left of the decimal point giving 26.4
12 Kg = 26.4 lbs

Let's try another, 79 Kg
79 X 0.2 = 15.8
Now treat 15.8 X 11 as 158 X 11
Remember back to page 35 how to multiply by 11.
1(1+5) and (5+8)8
1(6) and (13)8
The one in 13 carries over to the 6.
1(7) and (3)8
Bring them all together
158 X 11 = 1738
Remember that we were initially calculating 15.8 X 11 and how

many numbers are to the right of the decimal point in 15.8? One. So place a decimal point one place from the right of the answer 1738

15.8 X 11 = 173.8

CHECKING YOUR ANSWERS

You will be delighted to learn that there is a technique you can use to see if your answer is the right one or not.

Let's take a look at how we do this.

We have just calculated 17 X 22 and got the answer of 374

How do we know it is right?
Add up all the digits of our answer like so:
3 + 7 + 4 = 14 then add these digits, 1 + 4 = 5
So we keep adding the numbers until we get a single digit. In this case it is five.

So now we look at our sum, 17 X 22
Add each number of each part of the sum like so
1 + 7 and 2 + 2
giving us
8 and 4
Multiply these two numbers together giving 32
Add these two digits together, 3 + 2 = 5
Compare the two single digits; if they are the same you know your answer is correct. In this case we are right as we have 5 and 5.

Let's try 23 X 28

We get the answer of 644
So let's add all the digits up

6 + 4 + 4 = 14 and 1 + 4 = 5

Now look at the sum

23 X 28
2 + 3 X 2 + 8
5 X 10 = 50, 5 + 0 = 5
We get the same answer so our answer of 644 is correct.

SUMMARY SO FAR

I hope you have really enjoyed working through this book so far and also received something of value from it. The techniques you have learnt can be used in every day calculations and hopefully by now you can see the benefit of these mathematical techniques as opposed to your "usual" or normal methods of multiplication.

The next part of this book is the key to dividing.

Dividing by 9

Dividing by 9 has been seen by many as difficult to say the least. When dividing by nine a lot of people would divide by ten and use that answer as a rough guide when they are working with smaller numbers. This is fine as an estimate, but you will never get the correct answer this way.

But let me show you how to do it simply and effectively. What better way to demonstrate this than with an example.

9 / 71 is 9 divided by 71

I am now going to rewrite this sum:

9) 7/1

So what have I done here? Using the Vedic System to lay this sum out I have put the **divisor** (9) to the left of the sum with a bracket after it. This DOES NOT mean question 9 which 9) generally means.

So what is the 7/1 called and why is it written that way? The 7/1 is called the **dividend**. This is the number the **divisor**.

I have split the 7 1 into tens and units, 7/1

9)   7/1

What you do is bring down the tens unit (in this case 7), and put it under the single unit (in this case the 1).

9) 7/1
    /7

———

So can you see yet what needs to be done? What happens now is bring down the 7 from the tens units under the line, and then add the left hand side consisting of the 7 and 1…

9) 7/1
    /7

———
   7/8

So what does 7/8 mean? Well it means that 9 goes into 71 seven times with a remainder of 8. Remember, the / does not mean divide by, it is merely separating the tens and units.

Let's try another one.

9 divided by 53

Lay it out in the Vedic fashion…

9) 5/3

Write down five under the 3

9) 5/3
   /5

———

Bring the five down under the line from the 5/3 part and add the 3 and five together

9) 5/3
   /5

———
   5/8

Therefore 9 divides into 53 five times remainder 8.

What about this one…

9 into 81

Lay it out in the Vedic style…

9)   8/1

Write down 8 under the units

9)   8/1
       /8
    ‾‾‾‾

Bring down the 8 from the tens and add the 8 and 1 together

9)   8/1
       /8
    ‾‾‾‾
     8/9

So 9 goes into 81 eight times remainder 9. But is remainder 9 actually the correct answer? Well if we have a remainder of nine then we have a spare nine left over, as 9 divides into 9 once remainder 0. So in the case when we have a remainder the same as the divisor we add one to the left hand number, in this case add one to the 8 giving 9 and a remainder of zero…

9)   8/1
       /8
    ‾‾‾‾
     9/0

So nine goes into 81 nine times remainder nothing.

Let's try 9 into 99

9) 9/9

Write down the 9 from the tens under the units

```
9)  9/9
     /9
    ____
```

Bring down the first 9 under the line and add the two 9's on the right side

```
9)  9/9
     /9
    ____
    9/18
```

Nine goes into 18 twice with a zero remainder, so add 2 to the 9 giving 11

```
9)  9/9
     /9
    ____
    11/0
```

Nine goes into 99 eleven times no remainder.

You may have noticed how the remainder part can be determined by adding the two numbers from the dividend together.
The example of 9 goes into 72, add the 7 and 2 giving nine…
Or the 81 divided by 9, 8 + 1 = 9
Or 9 into 53, 5 + 3 = 8 (remainder 8)

So let's now try dividing 9 into numbers made up of more than two digits.

Take a look at 104 divided by 9.

Lay it out in the Vedic style as before:

9)   10/4

Now we don't bring down the 10 and put it under the four. We bring down the 1 and put it under the 0. So for all these examples we merely take the first number and place it under the second number:

9)   10/4
       1/

What now? Well just like dividing nine by two numbers where we added the numbers on the right hand side, we do the same here, but in this case we add the 0 and 1 together and place the answer under the 4 on the right hand side.

9)   10/4
       1/1
     ─────

Next we just bring all the numbers down under the line, remembering to add as we go:

9)   10/4
       1/1
     ─────
      11/5

What do I mean by add as we go? Look above. I brought down the first 1 and there was nothing under it so I didn't add anything to it. The next number, 0, I added to the 1 below it, and this gave me the 1 under the line. Finally I added the 4 and the one together giving me the five under the line.

Let's try another.

145 divided by 9.

9) 14/5

Bring down the one and put it under the four:

9) 14/5
    1/
   ─────

Add the 4 and one together and place under the 5

9) 14/5
    1/5
   ─────

Add the numbers in each column and place under the line,

9) 14/5
    1/5
   ─────
   15/10

So now we have a remainder of ten. 9 will go into ten once with one remainder. Leave the remainder to the right and bring the 1 over to the left hand side, adding it to 15.

9) 14/5
    1/5
   ─────
   16/1

Nine divides into 145 sixteen times with 1 remainder.

Let's try one more
Nine into 234

9) 23/4
    2/

———

9) 23/4
    2/5

———

Add and bring down under the line…

9) 23/4
    2/5

———

   25/9   (nine into 9 with no remainder…)

9) 23/4
    2/5

———

   26/0

9 goes into 234 26 times with no remainder.
What about dividing by numbers less than 9?

Dividing by 7

Let's look at dividing 7 into 46. We are going to use the same method as above except there is one slight adjustment to the calculation:

7) 4/6

When dividing by nine we notice that 9 is one less than ten. In this example 7 is three less than 10. We didn't need to include this part when dividing by nine as anything multiplied by 1 is the same.

7) 4/6
3 (this is the difference between 7 and ten).

7) 4/6
3
Next we multiply the 4 by the 3 and place it to the right hand side

7) 4/6
3   /12

7) 4/6
3   /12

———

Add the numbers and bring below the line:

7)   4/6
3   /12
     ──────
     4/18   (seven goes into 18 twice remainder 4)

7)   4/6
3   /12
     ──────
      6/4

Seven goes into 46 six times remainder 4.

Dividing by 8

How about 8 into 134

8) 13/4
2   (remember, 8 is two from 10)

Multiply the fist number of the dividend which is 1 by the 2 and place it under the next number, in this case 3.

8) 13/4
2   2/

Add the three and 2 and place under the 4 as we have done before:

8) 13/4
2   2/5
    ___

Add the numbers and bring under the line

8) 13/4
2   2/5
    ___
    15/9   (eight goes into 9 once, remainder 1)

8) 13/4
2   2/5
    ___
    16/1

Eight divides into 134 16 times with one remainder.

Simple Division

Division is deemed by many as more effort and harder to complete than multiplication.

Well let's eliminate that myth now with this instalment for division.

So, quickly let's do short division so as you can divide 6 into 76.

Layout the sum like so:

$$6 \overline{\smash{\big)}\ 76}$$

We need to ask ourselves what we need to multiply 6 by to give us 7? Well we can't multiply any whole number to give us seven, so we go for a number below 7. We can multiply 6 by 1 to give six and the difference between six and 7 is one (our remainder).

$$6 \overline{\smash{\big)}\ 7_1 6} \\ 1$$

So next we need to ask what we need to multiply six by to give us 16. Nothing will give us sixteen directly, so we try 3. 6 X 3 = 18. Too big.

6 X 2 = 12. This works.

So we place the two in the part of our answer. What's the remainder? Take twelve from sixteen giving us a remainder of 4.

$$6 \overline{\smash{)}7_1 6} \\ 12\ r\ 4$$

Let's try dividing 223 by 53

Lay out the sum like so:

$$5\ ^3 | 2\ 2 | 3$$

You will notice the 3 of 53 is slightly smaller. This **does not** mean 5 cubed, it is merely written this way so as you can see it is separate from the 5. We can call this our FLAGGED value.

So now we need to ask ourselves how many times does 5 (from 53) go into 22 (this first part of the sum, separated by those black lines).

5 goes into 22 four times with a remainder of two.

Note how the two remainder goes under the three.

$$5^3|22|3_2$$

$$\phantom{5^3|}|\phantom{22}4$$

Next we take that four and we multiply it by our flagged 3 that belongs to 53, giving us twelve.

So what next? Well we have our twelve which we just calculated, and we also have 23 from the remainder 2 and the three (above).

So we need to subtract 12 from 23 giving us 11.

Our answer now becomes:

$$5^3|22|3_2$$
$$\phantom{5^3|}|\phantom{2}4|11$$

4 remainder 11.

Pretty simple eh?!

Fractions

Whilst in school we all probably hated the thought of working with fractions. They seemed difficult, hard to grasp, and the techniques we learnt to calculate them was by no means easy.

You can rest assured the methods you are going to learn now make adding fractions a cinch.

Let's begin.

Adding Fractions

1/4 + 1/3

The method we use for adding **ANY** fraction is like so:

We cross multiply the fractions.

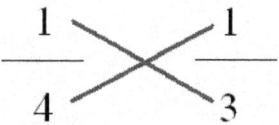

So what we have is 1 X 3 and 1 X 4

We add the results of both fractions together, so it will look like this:

(1 X 3) + (1 X 4) = 7

Next we need to find the denominator (the bottom part of the fraction). This is easily done by multiplying 4 X 3, giving 12.

1/4 + 1/3 = 7/12

Ok let's try 2/5 + 4/7

Remember same rule again we cross multiply

As before, (2 X 7) + (4 X 5) = 34

Now we need to find the denominator, to do this multiply the two bottom numbers together, 5 X 7 = 35

2/5 + 4/7 = 34/35

What about this one:
2/5 + 5/7

(2 X 7) + (5 X 5) = 39

Bottom denominator 5 X 7 = 35

2/5 + 4/7 = 39/35

But this isn't necessarily the fraction in its lowest form. We can see that the denominator is less than the numerator (39) and that it will go into 39 once with a remainder of four.

So how do we work it out this way? Well as it will go into 39 once we put a one at the start of the fraction and make the remainder and numerator into the fraction.

14/35

2/5 + 4/7 = 39/35 = 14/35

Subtracting Fractions

Subtracting fractions is just as simple as adding them.

Let's take the same example as one above and walk through it:

1/4 - 1/3

Again we cross multiply but instead of adding we subtract

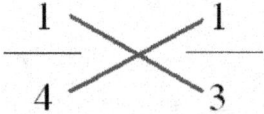

(1 X 3) – (1 X 4) = 3 – 4 = -1

Multiply the denominators together, (4 X 3) = 12

This gives us – 1/12

The whole fraction becomes minus 1 over twelve. Make sure not to make the mistake of saying the 1 is negative, the WHOLE fraction has become negative.

Multiplying Fractions

We will continue on from using previous examples, but instead of adding or subtracting we will multiply.

Ok let's try 2/5 X 4/7

$$\frac{2}{5} \quad\text{---}\quad \frac{4}{7}$$

There is a slight variation to multiplying fractions. Instead of cross multiplying, we just directly multiply, left side with right side. In the case above, we work it like so:

(2 X 4) / (5 X 7) = 8 / 35

Let's try another:

1/4 - 1/3

$$\frac{1}{4} \quad\text{---}\quad \frac{1}{3}$$

Remember, just like before, multiply top by top, bottom by bottom:

(1 X 1) / (4 X 3) = 1 / 12

Dividing Fractions

Welcome to the last week of our adventure through Vedic Maths.

By now you are probably thinking why you were not shown this method in school. Good question.

But anyway, on with the final instalment.

How to divide fractions.

Well if you can remember what we did for multiplying fractions, which entailed multiplying the numbers directly across from one another. Well you will be glad to hear that the same rule applies for dividing fractions, but with one small adjustment.

Ok let's try $2/5 \div 4/7$

Now, this sum can be laid out just like before when we were multiplying:

$$\frac{2 \longrightarrow 4}{5 \longrightarrow 7}$$

But what we do is swap the numbers around on one of the fraction and do as before for multiplying, so in this case it would be:

$$\frac{2}{5} \frac{7}{4}$$

Then we multiply across:

2 X 7 for the top

5 X 4 for the bottom giving us 14/20 as our answer, which if halved gives 7/10 (lowest this will go).

Square roots

The symbol for square roots is this: $\sqrt{}$

Imagine multiplying 25 X 25 and you get the answer of 625. Well the square root is the reverse.
$\sqrt{625} = 25$

So how do we go about performing this reverse function?

Well it's not that difficult at all!

Let's try $\sqrt{2209}$

We need to separate the numbers, giving us 22 and 09.

Look at the 22. We know that the first number MUST be 4 as 4 X 4 = 16.
5 X 5 = 25 and this will be above our value 22.

So what's our remainder? 4 X 4 = 16. Sixteen from 22 is 6, our remainder.

$\sqrt{22\ 6\ 09} = 4?$

To get the last part of the answer we divide 60 into TWICE the first part of the answer, in this case we use 4.

60 into twice 4 is 8 into 60 = 7.5

We can now drop the .5 from 7.5 as this can easily go back into the sum and cancel anyway, so we get the answer of 47.

$\sqrt{1089}$

Ok, we can see that the first part of the answer will be 3 as 3 X 3 = 9 which is below 10 by splitting 10 89 apart.

$\sqrt{1089}$= 3? There is a remainder of 1.

$\sqrt{10189}$= 3?
Twice 3 (6) then divide this 6 into 18, giving 3.

This is all the work you need to do! $\sqrt{1089} = 33$

$\sqrt{2809} = ?$

First part, 28. 5 X 5 = 25 (the first part of our answer).
25 from 28 is 3.

$\sqrt{28\ 3\ 09} = 5?$

Double 5 giving ten and divide this into 30, which goes in 3 times with a remainder of 0.

$\sqrt{28\ 3\ 09} = 5\ 3?$

$\sqrt{2809} = 53$

Estimating Square roots

√70 = ?

First let's try to guess the square root, from what we did before.

8 X 8 = 64.
9 X 9 = 81

So we know that the first part of our answer starts with 8.

Now we divide our 8 into the 70 giving us 8.75,

70 divided by 8 = 8.75

So now we split the difference between 8.75 and our 8 estimate, giving 0.75

Now split the difference of 0.75 giving us 0.375

Finally add this result to our estimate 8 giving 8.375

The answer will always be slightly higher than the correct one but for estimating purposes it is pretty accurate!

Let's go one more:

√29 = ?

Again take an estimate, 5 looks good (5 X 5 = 25).

Divide 25 into 29 and it goes once with remainder 4.
Five divides into 40 eight times. This gives us 5.8

Split the difference between 5 and 5.8 giving us 5.4

The correct answer is 5.385 but 5.4 is a pretty good estimate seeing as that's what we are aiming to do!

$\sqrt{31\ 25} = ?$

Pair off the numbers starting from the right,

31 25

Each pair has one digit answer to it.

Let's estimate the first digit of 31, 5 looks good. (5 X 5 = 25).

There is one more digit to the answer so we add one zero to our answer, 50.

To divide by 50 we divide by ten giving 5.
3125 divided by 10 is 312.5

We now divide by 5 giving us 62.5
So now we again split the difference 62.5 – 50 = 12.5
12.5 divided by 2 = 6.25

Round off downwards and add this to your first estimate of 50, giving 56

$\sqrt{3125} = 56$

Quick Sums

12 X 7

first thing is to always multiply the 1 of the twelve by the number we are multiplying by, in this case 7. So 1 X 7 = 7.

Multiply this 7 by 10 giving 70.
Now multiply the 7 by the 2 of twelve giving 14. Add this to 70 giving 84. Therefore 7 X 12 = 84! How quick was that!

Let's try another:

17 X 12

Remember, multiply the 17 by the 1 in 12 and multiply by 10 (just add a zero to the end): 1 X 17 = 17, multiplied by 10 giving 170.
Multiply 17 by 2 giving 34.
Add 34 to 170 giving 204.

So 17 X 12 = 204

Let's go one more

24 X 12

Multiply 24 X 1 = 24. Multiply by 10 giving 240.
Multiply 24 by 2 = 48. Add to 240 giving us 288

24 X 12 = 288 Simple!!

Is it divisible by four?

This little math trick will show you whether a number is divisible by four or not

So, this is how it works.

Let's look at 1234
Does 4 divide evenly into 1234?

Well for 4 to divide into any number we have to make sure that the last number is even. If it is an odd number, there is no way it will go in evenly.
So, for example, 4 will not go evenly into 1233 or 1235

So now we know that for 4 to divide evenly into any number the number has to end with an even number.

Back to the question...

4 into 1234

What you do is this:
Take the last number and add it to 2 times the second last number. If 4 goes evenly into this number then you know that 4 will go evenly into the whole number.
So
$4 + (2 \times 3) = 10$
4 goes into 10 two times with a remainder of 2 so it does not go in evenly.

Therefore 4 into 1234 does not go in completely.

Lets try 4 into 3436546

So, from our example, take the last number, 6 and add it to twice the penultimate number, 4

6 + (2 X 4) = 14

4 goes into 14 three times with two remainder.
So it doesn't go in evenly.

Let's try one more.

4 into 212334436

6 + (2 X 3) = 12

4 goes into 12 three times with 0 remainder.

Therefore 4 goes into 234436 evenly.

Is it divisible by 9?

Simply add all the digits up and see if they add to 9 or 18. i.e. 27 gives 2+7=9=9*3. 36 gives 3+6 = 9 = 9*4. 54, 45, 18, 81, 63. 11*9=99. 9+9=18. 56*9=504, 21*=189.

123 = 1 + 2 + 3 = 6, therefore 123 is not divisible by 9!

Conclusion

At this stage you have completed Mind for Math to help you get ahead with the basics of Multiplication and Division.

I hope you enjoyed your journey and discovered just how powerful some of these methods you have just learnt really are! In the near future I will be releasing a book on advanced tips and tricks to further your learning abilities with math to take you to the next level so be sure to keep a look out for my up and coming book.

All the best

Tim Perse

www.ingramcontent.com/pod-product-compliance
Lightning Source LLC
Chambersburg PA
CBHW051328220526
45468CB00004B/1551